I0797700

PINEAPPLES

by Emma Bassier

Cody Koala
An Imprint of Pop!
popbooksonline.com

abdobooks.com
Published by Pop!, a division of ABDO, PO Box 398166, Minneapolis, Minnesota 55439.

Printed in the United States of America, North Mankato, Minnesota.

052021
092021
THIS BOOK CONTAINS RECYCLED MATERIALS

Cover Photos: iStockphoto, foreground, background
Interior Photos: iStockphoto, 1 (foreground), 1 (background), 5, 7 (bottom left), 7 (bottom right), 8, 15, 21 (bottom left), 21 (bottom right); Red Line Editorial, 7 (top); Shutterstock Images, 9, 10, 13, 17, 21 (top); Marco Simoni/ Image Source/Alamy, 18

Editor: Aubrey Zalewski
Series Designers: Laura Graphenteen and Colleen McLaren

Library of Congress Control Number: 2020948861
Publisher's Cataloging-in-Publication Data
Names: Bassier, Emma, author.
Title: Pineapples / by Emma Bassier
Description: Minneapolis, Minnesota : Pop!, 2022 | Series: How foods grow | Includes online resources and index.
Identifiers: ISBN 9781532169823 (lib. bdg.) | ISBN 9781098240752 (ebook)
Subjects: LCSH: Pineapple--Juvenile literature. | Pineapple industry--Juvenile literature. | Fruit trees--Juvenile literature. | Agriculture--Juvenile literature. | Food crops--Juvenile literature.
Classification: DDC 631.5--dc23

Hello! My name is

Cody Koala

Pop open this book and you'll find QR codes like this one, loaded with information, so you can learn even more!

Scan this code* and others like it while you read, or visit the website below to make this book pop.

popbooksonline.com/pineapples

*Scanning QR codes requires a web-enabled smart device with a QR code reader app and a camera.

Table of Contents

Chapter 1

Sweet and Juicy

A pineapple is a fruit. The inside is yellow, sweet, and juicy. The outside is covered in **prickly** skin. A **crown** of green leaves sits on top.

Watch a video here!

Chapter 2

Growing Pineapples

Pineapples grow in tropical **climates**. These places are warm and wet. Pineapples grow in places such as South America and Costa Rica.

Where Pineapples Grow

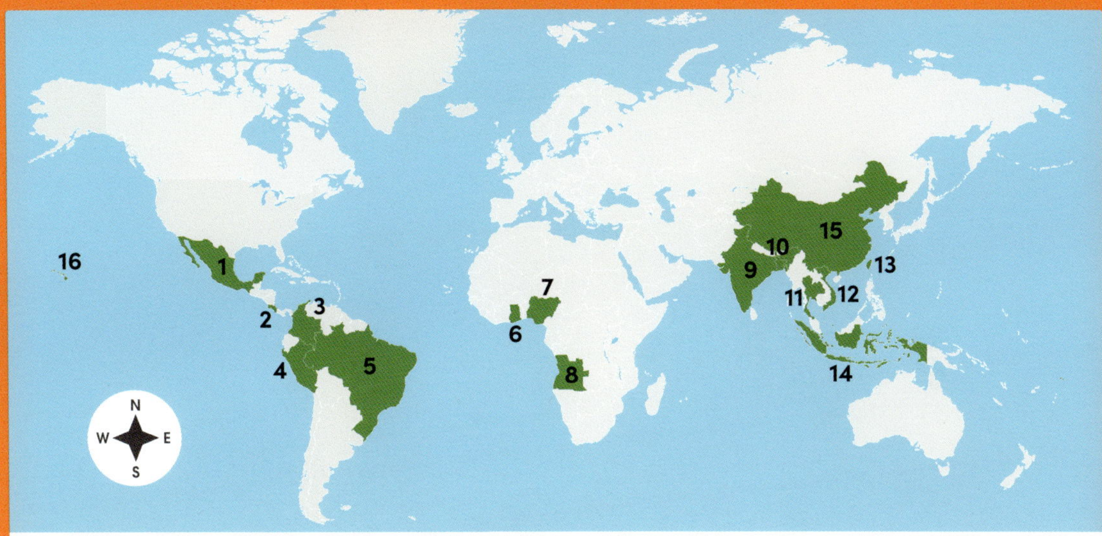

1. Mexico
2. Costa Rica
3. Colombia
4. Peru
5. Brazil
6. Ghana
7. Nigeria
8. Angola
9. India
10. Bangladesh
11. Thailand
12. Vietnam
13. Taiwan
14. Indonesia
15. China
16. Hawaii, United States

Learn more here!

A farmer cuts off a pineapple's **crown**. Small roots grow from the crown.

Then the farmer plants it. The roots pull **nutrients** from the soil.

A stem grows. The stem sends nutrients and water to all parts of the plant. Large, stiff leaves grow out of the top. They take in sunlight. Sunlight gives the plant energy to grow.

Pineapple plant leaves can grow to be as long as adults are tall.

Hundreds of flowers bloom from the stem. They are red and purple. The flowers grow many small fruits. The fruits form together. They become one pineapple fruit.

Chapter 3

Picking Pineapples

It can take pineapples up to 20 months to reach their full size. Then the fruit ripens after five to six more months.

Complete an activity here!

Pineapples change color as they ripen. Young pineapples are green and dark brown. Slowly the skin turns yellow and light brown. The **prickly** skin flattens. The fruit has a sweet smell.

leaves
crown
skin
fruit

Farmers **harvest** the ripe pineapples. Farmers pick them by hand. They cut the pineapples from the plant. Some farmers do this with **machetes**.

Pineapples stop ripening once they are picked.

Chapter 4

Cutting Pineapples

People must cut pineapples before they can eat them. People cut off the **crown** and skin. They cut out the tough center too. Then the pineapple is ready to eat!

Learn more here!

Making Connections

Text-to-Self

Would you want to see where pineapples grow? Why or why not?

Text-to-Text

Have you read other books about how foods grow? How were those foods like pineapples? How were they different?

Text-to-World

Pineapples grow in tropical places. What other foods grow in tropical places?

Glossary

climate – the typical weather of a place or area over time.

crown – the top of a pineapple, which is made of stiff, green leaves.

harvest – to gather or pick crops.

machete – a large, heavy knife.

nutrient – part of food that humans, animals, and plants need to stay strong and healthy.

prickly – covered in short, sharp points.

Index

Online Resources

popbooksonline.com

Thanks for reading this Cody Koala book!

Scan this code* and others like it in this book, or visit the website below to make this book pop!

popbooksonline.com/pineapples

*Scanning QR codes requires a web-enabled smart device with a QR code reader app and a camera.